Extreme Weather

DROUGHTS

Martha London

DiscoverRoo
An Imprint of Pop!
popbooksonline.com

abdobooks.com

Published by Pop!, a division of ABDO, PO Box 398166, Minneapolis, Minnesota 55439.

Printed in the United States of America, North Mankato, Minnesota.

052019
092019

THIS BOOK CONTAINS RECYCLED MATERIALS

Photographs ©: Marijan_Murat/picture-alliance/dpa/AP Images, cover, 1; Sakchai Lalit/AP Images, 4; Linh Pham/Getty Images News/Getty Images, 7, 29; Thai Navy/Polaris/Newscom, 9; Cpl. Adam Dublinske/US Marine Corps/Defense Visual Information Distribution Service, 10; Mass Communication Specialist 1st Class Abe McNatt/US Navy, 13; Senior Airman Ryan Conroy/US Air Force/Defense Visual Information Distribution Service, 14; aetb/iStockphoto, 17; Jorge Saenz/AP Images, 18–19, 23; Rott70/iStockphoto, 20; pedrosala/Shutterstock Images, 25 (left); Berenika_L/iStockphoto, 25 (right); Bai Yu/Xinhua/AP Images, 27

Editor: Connor Stratton
Series Designer: Sophie Geister-Jones

Library of Congress Control Number: 2018965056

Publisher's Cataloging-in-Publication Data

Names: London, Martha, author.

Title: Droughts / by Martha London.

Description: Minneapolis, Minnesota : Pop!, 2020 | Series: Extreme weather | Includes online resources and index.

Identifiers: ISBN 9781532163920 (lib. bdg.) | ISBN 9781532165368 (ebook)

Subjects: LCSH: Droughts--Juvenile literature. | Drought management--Juvenile literature. | Natural disasters--Juvenile literature.

Classification: DDC 363.34929--dc23

WELCOME TO DiscoverRoo!

Pop open this book and you'll find QR codes loaded with information, so you can learn even more!

Scan this code* and others like it while you read, or visit the website below to make this book pop!

popbooksonline.com/droughts

*Scanning QR codes requires a web-enabled smart device with a QR code reader app and a camera.

TABLE OF CONTENTS

CHAPTER 1

HOW DROUGHTS FORM

A drought happens when an area has less water than normal. One common cause of a drought is less rain. Too much **evaporation** can also cause a drought. Evaporation happens when the sun and heat cause water to dry up.

WATCH A VIDEO HERE!

During a drought, soil can become dry and cracked.

DID YOU KNOW?

When a drought makes cracks in the soil, buildings on the soil can also crack.

Changing weather patterns can also lead to droughts. For example, ocean water temperatures sometimes change.

Storms form when warm air and cool air meet.

Parts of the ocean may cool. Cooler water can cause fewer storms. Less rain from storms can lead to a drought.

The rising of Earth's average temperature is known as global warming.

Climate change is affecting droughts too. Humans have caused Earth's average temperature to rise. Many scientists believe this rise is making droughts worse. A warmer climate can increase evaporation. It also causes sea ice to melt. Melting sea ice could change where storms go. As a result, some places could experience more droughts.

Sea ice melts near Greenland.

CHAPTER 2

WHERE AND WHEN

A drought can occur anywhere. It is based on an area's normal supply of water. For example, a **rain forest** may get 80 inches of rain in an average year. If only 30 inches fall, a drought could start. In contrast, a **desert** may get only 10 inches on average. There, 30 inches would be much wetter than usual.

LEARN MORE HERE!

When pine trees do not have enough water, their needles can turn brown.

Some areas are more at risk of drought than others. In the United States, California faces a higher risk of drought than many other states. In Africa, droughts are more common south of the Sahara Desert. The soil there is not good at holding water.

DID YOU KNOW?

Desert plants are less affected by droughts than plants in other climates. The spikes of a cactus limit evaporation.

Much of California faced extreme drought from 2011 to 2017.

In 2015, a drought dried up many areas in the Amazon rain forest that were usually flooded.

Climate change can affect what areas are at risk for drought. For example, a severe drought hit the Amazon rain forest in 2005. Scientists thought that level of drought would be rare. But extreme droughts also hit the Amazon in 2010 and 2015.

In 2015, a drought in the Amazon trapped many boats at the bottoms of dried rivers.

Droughts end when rainfall returns to normal. Areas sometimes need months of rain to end a drought. And some droughts last for years. Scientists study weather patterns to figure out

One heavy rain may help plants survive a drought. But it's often not enough to end a drought.

when droughts may end. But these patterns are tricky to understand. Sometimes people can't tell when a drought started or ended until long after it is over.

DROUGHT IN THE UNITED STATES

JANUARY 2016

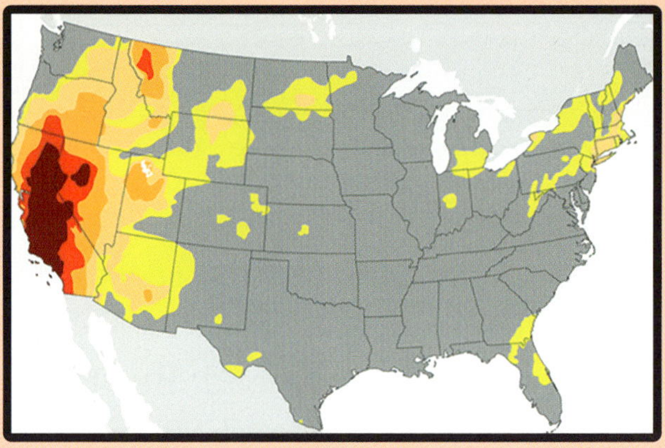

JANUARY 2017

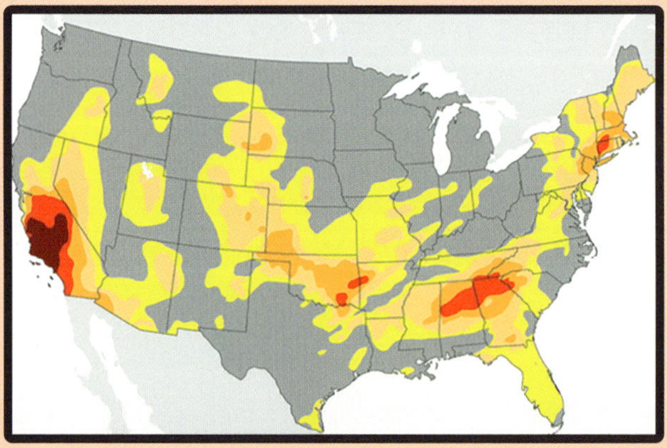

Scientists create maps to show which parts of the United States are having droughts. Different colors show how dry an area is.

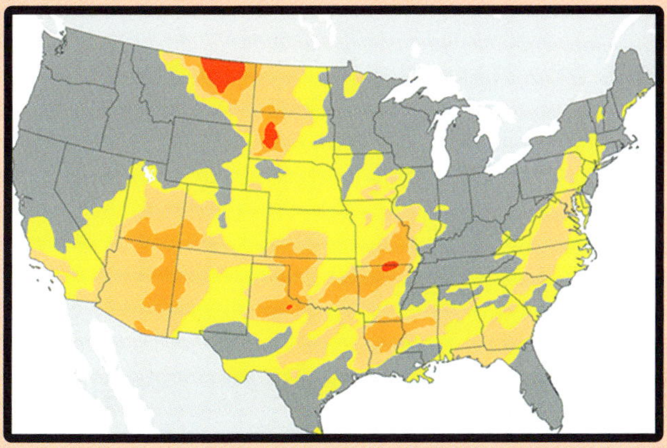

JANUARY 2018

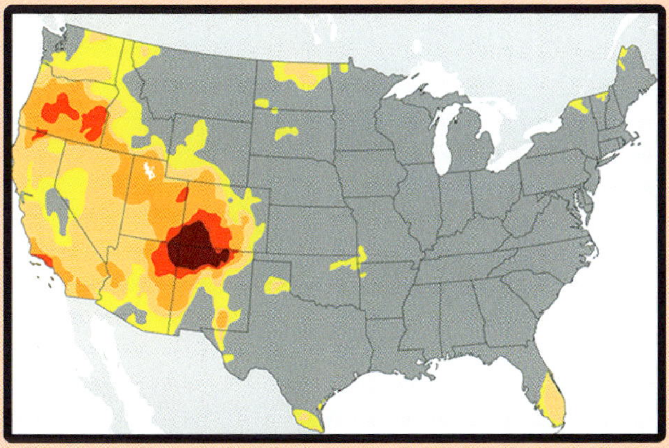

JANUARY 2019

CHAPTER 3
EFFECTS

Droughts are harmful in many ways.

Without enough water, crops can die.

If crops die, farm animals may die too.

These problems can also harm people.

They may not have enough food to eat.

DID YOU KNOW?

In the 1930s, a drought in the United States lasted nearly ten years. The area was known as the Dust Bowl.

The Dust Bowl got its name from the dust storms that formed from the drought.

LEARN MORE HERE!

Dry plants from a drought make great fuel for a fire.

Droughts also make **wildfires** more likely. All fires need oxygen, heat, and fuel. Dry fuel is easier to burn. During a drought, trees and plants get very dry. Fires can grow and spread more easily.

WORSE WILDFIRES

Between the 1970s and the 2010s, the average temperature in the western United States rose. This rise has increased the number of massive wildfires. Wildfire seasons are also becoming longer. In the 1970s, most wildfire seasons lasted five months. By the 2010s, the seasons sometimes lasted all year.

The Aral Sea in central Asia used to be one of the largest lakes in the world. But by 2000, the Aral Sea had shrunk.

In addition, **climate change** and drought can turn some areas into **deserts**. Droughts become a natural

By 2016, the Aral Sea had almost disappeared. Drought turned this area into a desert.

part of these places' climates. When an area becomes a desert, it is harder for people to live there.

CHAPTER 4

STAYING SAFE

During a drought, people use less water. That way, it will not run out. People take shorter showers. They water their lawns less often. Some people even buy sinks and toilets that **conserve** water.

COMPLETE AN ACTIVITY HERE!

Watering a lawn for one week can use as much water as all the showers a family takes in a year.

DID YOU KNOW?

An eight-minute shower uses 20 gallons of water.

Planting a tree can help an area for a long time. Some trees live for hundreds of years.

Other steps help save water too. For example, people plant trees. Tree shade cools the area around it. Cooler temperatures slow **evaporation**. All these actions help people have enough water until the drought ends.

Drought Safety Checklist

- ✓ Conserve water inside your home.
- ✓ Water lawns less often.
- ✓ Plant trees and shrubs.
- ✓ Follow the news for weather updates.

MAKING CONNECTIONS

TEXT-TO-SELF

Are droughts common where you live? Why or why not?

TEXT-TO-TEXT

Have you read books about other kinds of natural disasters? How are they similar to droughts? How are they different?

TEXT-TO-WORLD

During a drought, an area has less water than usual. Why does conserving water help during a drought?

GLOSSARY

climate change – differences in Earth's weather patterns, often including a rising temperature.

conserve – to save or use less of something.

desert – a dry area with little rainfall and few plants.

evaporation – the process of liquid water becoming water vapor.

rain forest – a warm, thick forest that receives a lot of rain.

wildfire – a large fire that grows quickly because of very dry plants and trees.

INDEX

ONLINE RESOURCES

popbooksonline.com

Scan this code* and others like it while you read, or visit the website below to make this book pop!

popbooksonline.com/droughts

*Scanning QR codes requires a web-enabled smart device with a QR code reader app and a camera.